Why do Golf Balls

have Dimples?

Published by Accent Press Ltd – 2012

ISBN 9781908192905

The Quick Reads project in Wales is a joint venture between the Welsh Government and the Welsh Books Council.

Printed and bound by CPI Group (UK) Ltd, Croydon, CR0 4YY

Illustrations by Liz Bryan Graphics
Cover design by Madamadari

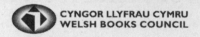

CYNGOR LLYFRAU CYMRU
WELSH BOOKS COUNCIL

Noddir gan
Lywodraeth Cymru
Sponsored by
Welsh Government

Why do Golf Balls have Dimples?

And other weird and wonderful facts...

Wendy Sadler

ACCENT PRESS LTD

Chapter One

SPORT

Can bungee jumping make your eyes fall out?

Bungee jumping is surprisingly safe, with only a handful of deaths reported since the sport began in 1979. But damage to the body is pretty likely because of the huge forces on you when you reach the bottom of the jump. In a bungee jump you're falling at very high speed and then you're stopped and pulled back up very suddenly by the elastic reaching the end of its stretch. This change in direction makes some parts of your body want to carry on moving down, whilst others are pulled back up by the bungee rope. Sounds painful already...

Now for the good news. Your eyes are fixed into your skull by six muscles that make your eye move up and down and left and right. Your eyeball also fits very tightly into the socket of your skull. So it is pretty difficult to dislodge an eye completely – even by sneezing with your eyes open (but that is another story!).

There have been tales of people losing their eyesight after a bungee jump, though – so what's happening? Well, when you reach the lowest part of the jump the blood in your body gets pushed down to the bottom end of you all at once. As you are upside down, this means blood rushes to your head and your blood pressure goes up a lot. This can have a few nasty side-effects that you might want to consider before leaping from that bridge.

There is a part of your eye called the retina which you use for seeing colours and light and dark. If the blood pressure in your eye gets very high it can make this retina come loose, and a loose retina can mean that the signals from your eye to your brain get mixed up. You may start to see flashes of light or strange floating shadows in front of your eyes. This can usually be put right if you get to a doctor quickly enough, but if not it can give you permanent damage. So in short, your eyes won't fall out, but, if you're unlucky, they may never be the same again.

Why do golf balls have dimples?

In the beginning, golf balls were smooth. But then golfers began to notice that the older and rougher the balls got, the further they would fly. The old scuffed-up balls were in big demand, though no-one really understood what was going on.

Common sense would suggest that smooth things should cut through the air more easily. Cars and boats that are designed to go as fast as possible usually have smooth surfaces and streamlined shapes. The force that slows things down in air is called drag, so how come a rough ball would give you *less* drag?

When a smooth ball travels through the air, the air flowing around it doesn't meet up nicely at the back of the ball. The path of air over the top and the path underneath split up and create an area of turbulence behind the ball that slows it down. But if you take a ball with a rough surface, the air flowing around it sticks to the ball longer until the air almost meets up again neatly at the back. The fact that the air doesn't split so early on means there is much less drag at the back of the ball. This in turn means it isn't held back and flies much further.

Fig 1. Flow over a
smooth ball

Fig 2. Flow over a
dimpled ball

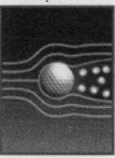

The shape, pattern and depth of the dimples can make a difference, too. Most golf balls have round dimples, but some now use hexagon shapes which make the drag even less. The dimples are usually arranged in a pattern that is the same all over the ball, but in the 1970s one

4

company made golf balls with a special design where the dimples were deeper in some parts of the ball than in others. This helped the ball correct itself in flight. These were banned from use in competitions but can still be bought for private use if you need a bit of help on the golf course! If you ever find yourself with too much time on your hands you could try counting the dimples in your golf ball to see where it was made. American ones have 336 and British ones have 330!

Is it safe to sneeze when you are driving?

You know the dreaded burning feeling up your nose which tells you a sneeze is about to happen? And the more you think about it, the less likely it is to go away? Add this to the fact that you are driving down the motorway at top speed. Are you about to put your life (and other people's) at risk?

Let's work out what might happen. Although any distraction can be dangerous, the biggest problem with sneezing is that your eyes will be closed – generally not a good idea when driving! If we guess that a sneeze lasts about one second we can work out how far you will actually travel when going 70 mph and it's about 31 metres (or 101 feet).

This is about six car-lengths, so certainly enough of a distance to cause problems if you are on a fast-moving road, or something changes unexpectedly. What makes accidents less likely is that you can prepare for a sneeze. You usually get a good few seconds warning and this lets your brain take into account what is happening around you to help you prepare.

So, to prevent any danger, could you train yourself to keep your eyes open when you sneeze? Unlike what you may have heard, your eyeballs will not pop out if you sneeze with your eyes open. Scientists don't fully understand why we close our eyes when we sneeze. It isn't to protect our eyes from the germs coming out of our noses, as they are fired out at high speeds of up to 100 mph! A few people can keep their eyes open when sneezing but for most of us the blinking is what is known as a reflex action. We just can't help it. The same type of thing happens when you tap below your knee and your leg kicks up. It just does. However hard you try, it isn't something you can learn.

The other thing you could do if you feel a sneeze coming on, is to press firmly either just under your eyebrow, or against your top lip. The signals that go from your nose to your brain pass by both these areas of your face. If you do it quickly enough you may be able to prevent the sneeze for a little while, perhaps even long enough to pull over somewhere safe...

Why do you spin a rugby ball?

When rugby began as a sport they used a pig's bladder wrapped in leather as a ball. The bladder had to be inflated by someone blowing into it with their mouth – not the most popular job in town! The shape of the ball is a really important part of the sport. Compared to a round ball, it is easier to carry under the arm, easier to pass and because it doesn't roll as far the ball stays in play for longer. But there is another advantage to the shape that, when used correctly, can help make you a better player.

One problem with an oval shape is that when you throw it through the air, any wobble that it has makes a lot of drag, which slows the ball down. If you watch good rugby players throw a ball, you will see that they twist their hands as they throw. This twist makes the ball spin like a corkscrew as it flies through the air. The idea of the spin-throw is to stop the ball wobbling so that the ball can travel a lot further. But why does the spin make the ball more stable in the air?

If you've ever been on a bike you will have noticed that it is difficult to stay upright when the bike is going slowly but as you go faster it gets easier. This is because of something called the gyroscope effect. The spinning of the bike wheels create a force that makes the bike less likely to fall over. You may have seen small gyroscope toys that can balance on your finger or on a tiny point for as long as they keep spinning, but as soon as they stop they fall over.

The spinning of the rugby ball does the same thing. It keeps the ball lined up in the same direction as it flies through the air, and with less wobbling you get less drag, so the ball has more chance of making it over those all-important bars.

Where is the best place to aim in ten-pin bowling?

The best result you can hope for in a round of ten-pin bowling is a strike, where all ten pins are knocked down by just one ball. In fact, a perfect game of bowling, where you score a strike every time, will score you 300 points.

So what do you need to know to get the best chance of achieving the perfect score?

The pins in bowling are arranged in a triangle shape with one pin (which is known as the 'head pin') at the point nearest to you. Common sense might tell you that aiming for the head pin will give you the best chance of a strike. But is this true?

If you hit the head pin straight on it is likely that some pins will stay standing. Worse than that, they are likely to be left at the opposite sides of the triangle, where you are very unlikely to hit them all with a second ball. This is known as a 'split'. For the best chance of a strike, right-handed players should aim the ball at the gap between the head pin and the one on its right. The ball should make direct contact with at least four of the pins to get a strike, and this is more likely if you aim for this gap. Left-handed players should aim for the gap to the left of the head pin. In both cases this area is known as 'the pocket'.

In an ideal throw, your bowling ball should skid along the first bit of lane, then start to roll and then finally bend around to hit the pins in the last part of the roll. Bowling lanes are oiled to encourage skidding but you need some friction on the lane further down to get the ball to start rolling. When the ball starts rolling, any spin you have put on the ball starts to have an effect and this steers the ball – with luck towards the pins.

Bowling used to be played with nine pins set out in a diamond shape, but the game became linked with crime and gambling and was banned. The extra pin was added to get around the ban so people could carry on playing!

Do you need two eyes to be good at sport?

It depends on the sport, but it certainly helps. Humans (and all mammals) have two eyes to help them to judge distances and to see the world in three dimensions. In times gone, by this would have been important when you were hunting food, or perhaps running away from something that wanted you as food (without bumping into things as you went!).

In most sports, judging distances from you is pretty important. You might need to know how far away a ball is so that you can hit it successfully with your bat, or see how far away the goal is to judge how hard you need to kick. It's quite hard to think of any sport where you don't need to be good at working out distances for one reason or another.

When you look at the world with two eyes your brain is adding up the images that both eyes see. Because your eyes are at different positions in your face, you see things from a slightly different angle through your left and right eye. Your brain puts these signals together to tell you how far away something is.

You can see this for yourself if you put your hand in front of you with one finger pointing up, and then try and bring your other hand over your head to touch the finger. It should be quite easy with both eyes open but if you close one eye and try it again, you'll find it much harder. This shows you that judging distances using only one eye can be quite a problem. If you have only one eye, you can practise this and get better at it, but it definitely puts you at a disadvantage compared to those with two good eyes.

Gordon Banks, who is commonly known as the best goalkeeper to have ever lived, had to cut his football career short in 1972 when he lost the sight in one eye in a car crash. Judging distance is pretty vital when you are in goal at the World Cup! Red Pollard, on the other hand, was a US jockey who famously rode a horse named Seabiscuit. He lost his sight in one eye but went on to become a champion jockey.

11

Horse racing is one of the few sports where judging distances is less important, as you have another pair of eyes on hand to help you!

Why do some Formula 1 tyres have no tread on them?

When you take the car for an MOT, one of the things they have to check for is the depth of tread left on your tyres. Anything below 1.6 millimetres is not allowed and can affect your breaking distance.

So why are some Formula 1 car tyres completely smooth?
The job of a tyre is to grip on to the road surface. Because the tyre is made of a very 'grippy' material there is a lot of friction between the tyre and the road. Friction is a very useful force that can slow you down and stop you slipping all over the place. Some Formula 1 cars use completely smooth tyres in dry weather. A completely smooth tyre has lots of contact between the rubber and the road, so these smooth tyres give great grip for fast corners.

When you're driving around in your hatchback, you can't pull in at a pit-stop to have your tyres changed when the clouds open, so normal cars need tyres that can grip in all weathers. The tread patterns on the rubber do not help with grip, but are there to channel water off the road. If you let your tread get too shallow, the water can't go anywhere and the wheels can skim along over the surface of the water. This is called 'aquaplaning' and is nasty for a car on any road, but could be much worse for Formula 1 drivers because of the speeds that they drive. So they can choose from a whole range of tyres to use during a race, depending on the weather. Dry weather races are usually faster and more exciting because of the extra grip the smooth tyres can give on the race track.

Why does sweat smell?

There's nothing like the smell of a changing room on a hot day! Sweating is a good way for the body to cool itself down – but does it have to smell so bad?

Your body likes to work at a certain temperature range. If you get too hot, a signal from your brain tells things called sweat glands to release sweat. This sweat is almost all water with a little bit of salt. If you have ever been swimming on a windy day, you will know that when water gets on your skin and you add in moving air, it can feel very cold. The wind blowing turns the water on your skin from a liquid into a gas and as this happens it takes energy from your skin and cools you down. Sweating does exactly the same job. The sweat glands that cool you down are called 'eccrine' sweat glands.

But if we all sweated just pure water, the deodorant market would disappear overnight, so something else must be going on.

There is a different type of sweat gland that isn't there just to cool you down. In certain parts of the body (the armpit, the ears, the eyelids and the belly button!) there are different glands that send out sweat that also has proteins and oil in it. And even though you may scrub yourself in the shower every day, you are actually covered in little bacteria. These bugs love to eat the tasty protein and oil in your sweat. But just like us they produce waste when they eat, and it is this waste that makes that sweaty smell. These type of glands are called 'apocrine' glands.

Vegetarians usually have sweeter-smelling armpits than meat-eaters because the meat-eaters have more protein in their sweat to feed up those hungry bugs. Chinese people don't have as many of the smelly type of sweat gland as white or black people, but no-one really understands why!

How can a mouthguard protect against brain damage?

If you ask someone which part of the body a mouthguard protects, they may well look at you as if you are mad! A mouthguard does of course stop you from hurting your teeth, mouth and tongue, but it can also help you avoid an injury to the brain.

When you are playing sport an injury can be caused by the force of a ball, or the ground, on different parts of your body. All types of protective sports clothes are designed to cushion the force that might be pushing against bits of your body. The impact on your body is worse if all the force happens at a very small point, or over a very short time. What padding does is spread out the force (in terms of both distance and time) so that the impact on your body is less. But how does this apply to your brain and your mouth?

Put your fingers up to your ears and rest them gently on the part where your jawbone meets your skull. Now just open your mouth by about a centimetre. You should feel a small gap open up where your fingers are. Wearing a mouthguard forces your teeth apart like this, which then opens up this small space between your jawbone and your skull.

If you were to fall on the floor or get hit on the chin with your teeth closed, all the force goes up into your skull and makes your brain shake around, possibly causing brain damage. By wearing a mouthguard, you can cushion this force and spread it out so that it is not all passed on to the skull.

Why do you get cramp?

The short answer is that scientists don't really know. We know what is happening but there is some confusion about why. Long-distance runners and tennis players are very likely to get cramp as it is often linked to one muscle being used heavily for a long period of time.

Cramp in your leg when you are asleep is the most common type for non-athletes to have and can feel like your leg muscles have locked up. Some people think this is because when you sleep you tend to have your legs flat with your toes pointing downwards, especially if you like your blankets tucked in very tightly! In this position, your calf muscle is shortened or contracted. For some reason, this means the muscle is more likely to go into a sudden spasm and contract even further, which is incredibly painful. The best thing to do is to gently try to stretch the muscle out by pointing your toes to the ceiling and rubbing the muscle to try to relax it.

Experiments have shown that if your body is dehydrated, or short of water, then cramps are more likely to happen. Being low on some vitamins and minerals also seems to have an effect. A banana is a good way to get a bit more potassium in your body and that can help prevent cramps. In bed you can help by raising your legs slightly with a pillow or by doing regular calf-stretching before sleep.

Chapter Two

ENTERTAINMENT AND GOING OUT

How do you fit all that music on an iPod?

When music first came out on CD in the 1980s the world was amazed. How could you fit all the music from two sides of a 12-inch record onto something so tiny? And could you really drive a car over it and still play it afterwards?! Fast-forward thirty years and even CD collections are huge in comparison to the thousands of albums that can fit on one iPod.

So how have we managed to squash all the music up into such a small space?
The main difference between the way we used to enjoy music and the way we do now is that it has less information in it. That sounds pretty weird – after all it sounds the same as, or perhaps even better than, the old-style vinyl records.

Music is just a collection of different sound waves all happening at just the right time. When you record music a microphone turns the

19

sound waves into shapes. The old-style vinyl records used to have the sound waves actually cut out of the surface by a needle. To play the music back, another needle went into the groove and turned the shapes back into sounds that came out of your speakers. All the information from the recording was kept, and you heard every little bit back. All good until the dust got into the groove or your cat walked over it and scratched the shapes around a bit!

When electronic gadgets and computers came along, information began to be recorded in a different way. The sound wave is measured and a code of numbers is recorded onto the CD. A laser beam inside your player then reads the code and turns the numbers back into sound.

If I carried all the music I have on my iPod as vinyl records it would weigh about the same as two men, or one dolphin!

How do 3D films work?

Most people see the world in pretty good 3D, unless you only have one eye, or, like me, have very poor eyesight in one eye. You can see how this works by doing a very simple test. Pick an object in the distance somewhere, look at it with

both eyes, then close one eye at a time and see how it looks. You should see the object looking slightly different from each eye. Your brain adds up both these versions of the world and it means you can see things in 3D and can work out how far away things are.

If you're old enough you may remember some of the first attempts at 3D films in the 1950s that were black and white, which you had to wear groovy red and blue specs to watch. Any 3D film has to be filmed with cameras that have two lenses (a bit like your eyes) that film the action from two different points of view. When you play the film back you then need some way of telling your eye which part of the film is coming from the right, and which from the left. The red and blue glasses allowed only blue light through the one eye and red light through the other. If you ever took the glasses off you would see a mess of red and blue shadows all over your monster (they were usually monster movies!). The problem with this was that it was very difficult to use in colour films, so it didn't become very popular.

The latest way to make 3D films uses the same idea but without the red and blue glasses. It uses a nifty trick of light called polarizing. When you go to the cinema or watch TV you are being bathed in light of different colours. Light is a kind of wave and usually it vibrates in lots of different directions all at once. This is known as unpolarized light, but if you take a polarized filter (like you might find in a pair of sunglasses) you can block out all the light except the bit that is vibrating in just one direction. Current 3D films use special polarized glasses where in the left lens you allow light vibrating in one way through, and in the right lens you let light vibrating in another way through.

Using this type of 3D means that you can still have full colour films, or even computer generated films (CGI) where a computer works out what would be seen by each eye if the object really did exist. Weird...

Unfortunately for people like me (with one very weak eye) this effect is not very impressive because my brain has got used to getting most of the information about the world through just one eye. Guess I'll stick to the 2D films then...

How does a Nintendo Wii know where my hands are?

Knowing where your hands are is the fairly simple bit. Knowing what they are doing and where they are going is a bit more difficult. To know where your hands are, each end of the Wii controller gives out an infra-red light beam. Infra-red light is invisible to us but it is the same type of light as your TV remote uses. A sensor bar that you place on top of your TV picks up both these signals and that tells the game where the controller is being held.

But what about how fast you are moving, or in which direction? Well, inside the controller are things called accelerometers, which as you might guess are used to measure acceleration, or change in speed. You can imagine them as little springs holding weights. Some are fitted to sense up and down movement, some side to side and some forwards and backwards. By adding up all the moves made in three directions, the games machine can work out how you are moving the controller.

By standing on a balance board you get another level of information about what your body is doing. The balance board has four sensors, one at each corner and each one measures the weight pushing down. By using information from all four sensors, the board can work out how you are standing or moving on the board. The inventors of the balance board got the idea for adding sensors from how sumo wrestlers weigh themselves using two sets of scales.

Your normal bathroom scales at home have just one sensor, so a balance board is much more accurate (and fun!).

Why does the curtain stick to me when I turn the shower on?

You know the frustrating feeling that someone is tickling you on the back just as you get lathered up in the shower? You turn around and end up all wrapped up in your plastic shower curtain. Why can't your shower curtain leave you alone?!

We don't usually give much thought to the air and what it is doing because we can't see it. Throughout our day-to-day lives there is a huge weight of air pushing against us. We are completely 'under pressure'. But sometimes this air pressure can be changed if the air is moving very fast, or changing in temperature.

In a shower, you aren't moving very fast but the water around you is. The water spray sweeps air along with it so you are standing in fast-flowing air and water. Before you turned the shower on there was some air outside the shower pushing against the curtain and some inside the shower pushing it back. Because there was a push from both sides you don't see anything happen, like a very even arm-wrestle where no-one moves.

When you switch the shower on, the air inside is now moving quickly downwards and this means it can't push out to the sides as much as it did before. This makes an area of lower pressure inside the shower by your body. The air outside the shower is still pushing by the same amount as it was before but the air inside isn't. Imagine the arm-wrestle situation and one person has just been tickled and can't push as hard any more. The air outside pushes in towards you and the shower curtain is in the way, so it gets pushed towards you as well.

Someone has invented a special shower rail that curves outside the shower space at the top to help stop this happening but it is not clear if it works. Solution? Buy a glass shower door...!

How does fake tan work?

A tan used to be the sign of the working class. Those with money would stay indoors when it was too hot (or cold) but these days a tan is most definitely a badge of the rich and famous. But with all the worry of skin cancer, faking it has become the next best thing.

There are three basic ways you can fake a tan without seeing the sun...

1. Use make-up (or bronzer if you're a bloke!).
2. Use spray-on or fake tan.
3. Use tanning pills.

The make-up option is simple. You just rub in a cream or powder that has brown colouring in it. When you've had enough (or when it rains suddenly) the colour can be washed off.

The second option is the most popular and there are hundreds to choose from. All fake tan contains a chemical called DHA which is a kind of sugar. The DHA reacts with the layer of dead skin on top of your body and changes its colour. As this happens you usually start to smell a bit like a sweet biscuit! This is the smell of the stuff doing its job on your dead skin. Nice!

Your dead skin cells will fall off over the next 5–7 days, which is why fake tan of this type can never last very long and sometimes looks a bit patchy as it fades.

The final way to change the colour of your skin is to take a pill that contains a browny-red food additive which is also used in barbecue and tomato sauces. Some vegetables, like carrots, contain stuff that can change the colour of your skin but you have to eat them in massive amounts to make any difference. The chemicals in tanning pills don't only colour your skin, but also your liver and sometimes your eyes too! Although this ingredient is safe to use as a food additive it takes much higher amounts of it to affect your skin, so it hasn't been shown to be safe. Brown skin might be a good look, but yellow eyes? Maybe not.

Why do my ears ring after a loud party?

You know the feeling – just when you're ready for the party to end, a whole new party starts in your ears. Long after the DJ has packed up and gone home, and sometimes even as the sun rises the next day, you can still hear ghostly sounds in your head. The good news is, it means your ears are working the way they should. The bad news is, it could mean you have caused some damage.

Sound is a wave of changing pressure through the air. Very loud sounds (like you would get at a live gig or nightclub) are a big change in pressure. As the sound wave goes into your ear it makes your eardrum move in and out, and connected to the eardrum are three tiny bones – the smallest in the human body. They pass on the sound to the most important bit of your ear – the inner ear. This is where the sound stops being a physical movement of parts of your body, and turns into an electrical signal to your brain. But this is also where most types of hearing loss happen.

The inner ear has lots of tiny hairs inside and when they move around they trigger a message to the brain to tell you what you are hearing. But if you put your ears in a very loud place for a very long time, the hairs can get worn out and then they can start constantly sending signals to your brain – even when the sound has stopped. It's a bit like some inner switch has got stuck in the 'on' position. This is what causes the whistling or ringing in your ears called 'tinnitus', and it gets more common as you grow older.

Experts worry that in the future there may be a whole generation of people with this problem because so many of us now listen to music through headphones at a very loud volume. As there isn't a cure yet it's probably worth turning the music down, or at least giving your ears a break for some time each day.

Is bacon good for a hangover?

Some people call it 'kill or cure'. Is a greasy fry-up of any kind going to help that morning-after feeling? And what about something for the poor vegetarians who like a drink or two?

To work out what's best we need to understand what's happening when you have a hangover. When alcohol gets into your bloodstream your brain tells your kidneys to send any water straight to your bladder. In fact, up to four times as much liquid can be lost compared to the amount you have drunk. This means all parts of your body get less water than normal and you get dehydrated. Your body then tries to steal water from wherever it can, and usually the water from the brain goes first. As the brain loses water it shrinks in size and pulls on the parts that connect it to your skull. No wonder your head is now pounding!

All this loss of water from regular trips to the toilet means your body also loses lots of important vitamins and minerals, and this is what makes you feel so sick and tired.

So how can food help, and what should you choose?
Eggs and bananas are a good start – but maybe not together! Eggs contain a chemical that helps break down the toxins of the alcohol and bananas contain a lot of potassium to replace what you have lost on your toilet trips. Anything that adds sugar or carbohydrate to your body, like fruit juice or toast, can also help with your energy levels.

So overall bacon is OK if you fancy it, but it doesn't really have any magic ingredients. Bacon contains protein, which is helpful, but it would actually be more helpful if you'd eaten it *before* you started drinking to line your stomach.

So, next time you are suffering, bring on the banana and egg butty. That'll sort you out.

Chapter Three

THE GREAT OUTDOORS

Why does hair go frizzy in the rain?

Not all hair does this, but if your hair has any wave or curl to it you may well have had this experience. You go out looking glossy and groomed but after the smallest amount of drizzle, you hair is ruined. Even a warm humid day can have the same effect. But there are ways to prevent the frizz.

Your hair is made up of three parts. On the outside of each hair is something called the 'cuticle' which if you looked at it under a very strong microscope would look like lots of roof tiles overlapping. On healthy hairs these lie very smoothly but as hair gets damaged (from chemicals or by brushing too roughly) these 'roof tiles' can get damaged and things in the air can get inside. The biggest part of a hair is called the cortex, and the shape and structure of this decides whether you have curly or straight hair. A round cortex gives you straight hair and an oval cross-section makes your hair curly or wavy. Inside the cortex are things called bonds that hold the different materials together. Some of these bonds are fixed and can't be changed but some of them can be changed by water.

When you spend a lot of time using heat (from a hairdryer or straighteners) to change the shape of your hair, you are breaking some of these bonds, but they always remember the shape they want to be in. Going out on a damp day lets water get inside your hair cortex and this can re-set the hair to its natural shape by changing the way the bonds are connected. If you have naturally straight hair, it will get straighter and might seem limp. If you have naturally wavy or curly hair, it will get more curly.

The best way to protect against this happening is to coat the hair with something that stops water getting into the hair. Anything containing silicon is a good bet for this. Or just wear a hat!

How lost can I get with a satnav?

As long as you use a bit of common sense, and your eyes (and your batteries don't run out!), not very lost at all. All satnav gadgets use a set of satellites above the Earth known as the Global

Positioning System (or GPS for short). Way back when sailors were exploring the world, they used stars in the sky to navigate. These satellites we use are the modern equivalent, but much more accurate and better on a cloudy night! There are 24 GPS satellites orbiting the Earth in such a way that they seem to stay in the same place in our sky. As long as your gadget can receive signals coming from at least three of them, they can tell you where you are.

Think of it like this – if you are trying to describe to someone on the phone where you are in a supermarket, you have to use things you can see. If you only use one thing like "I can see the freezer counter" it doesn't help much. If you add another two things like "the shampoo shelf is behind me and the loo rolls are on my right" you can pin down your position pretty exactly.

The satellites are fitted with very accurate clocks called atomic clocks. They send out signals all the time and your satnav receives these signals. Each signal is stamped with a time, and when you have received times from three different satellites the receiver can do a simple sum to work out your distance from each satellite. It can do this because the signal travels at the speed of light (which we know) and we also know what time the signal set off. It's like knowing that your friend set off from their house at 5 pm and was travelling at 50 mph – as soon as you know what time they arrive, you can work out the distance they have travelled. If they get to you after an hour, you know they live 50 miles away, if it only takes half an hour they were 25 miles away, and so on.

Although the satellites are over 11,000 miles away from Earth, the light travels stupidly fast and the clocks on the satellites are very accurate. The biggest error comes from the clocks in our satnav which are not quite so good. Military GPS can pinpoint your location within 5 centimetres but most of us can expect to get within 15 metres, on average. Not bad compared to the old nightmare of reading a map whilst driving round a city on your own – and you get someone to shout at too!

Why do stars twinkle?

It's something we've known from an early age thanks to the nursery rhyme, but have you ever wondered why? Most people have spent at least one evening of their lives looking up at the stars and wondering about big questions and the meaning of life. So let's answer at least one of the questions that may have come to mind on that occasion.

Whichever stars you are looking at, you can be sure that you are looking at light coming from something that is very, very far away. Our nearest star is the sun, which is so bright we don't see any other stars when it is out in the day. Even so, it would take about 177 years non-stop driving to get to the sun in an average car – and this is our closest star by a long way! Because of the big distances involved, we measure things in space in *light-years*. A light-year is a massive distance because it is how far light could travel in a year. A bit mind-bending but we have to use it or we run out of zeros for the numbers we need!

The next nearest star to Earth is called Alpha Centauri and it is four light-years away. Most of the stars you look at on a clear night are much further away than this.

But why should the distance make a difference to how the stars appear to us?
Light coming from the stars travels in a straight line and because the stars are so far away you can treat them as if they are just one single point of light. The twinkling happens because this single ray of light has to travel through the

Earth's atmosphere to reach us. In our atmosphere there are lots of different temperatures moving around and the light gets bent more by cold air than warm air. Because the light gets bent in and out of your eyes (as if it is switching on and off very fast) you see this effect as twinkling.

Not everything you see in the night sky twinkles, so why does the light coming from the moon or a planet not have the same effect? The light coming from the moon is coming from a very large area compared to the stars. The twinkling effect still happens but because there are so many rays of light coming into your eyes from all across the moon, you don't notice it. If you have ever seen a planet in the night sky, they seem to be points of light and can be mistaken for stars, but they are still large enough to be sending more rays of light into your eyes than a distant star.

So next time you lie on your back on a warm summer evening gazing at the stars, anything that doesn't twinkle is actually a planet, and if it moves across the sky, then it is most likely a satellite – or a plane!

Can you get sunburnt through a window?

Glass is a funny substance when you stop to think about it. It is made from the same basic stuff that makes up sand (called silica) but it is completely clear. Why does light travel so easily through the glass and can the dangerous bits get through too?

Glass is transparent to light because the way the material is made means that the energy in the light does not get completely used up, or absorbed, as it goes through. The smooth surfaces also mean that there are no reflections or bending of the light.

Light isn't just what we can see. Beyond the colours of the rainbow there is lots of stuff going on that is invisible to us. Past the red end, you get infra-red that we use in our TV remote controls and night vision cameras. If you go to the other end you get ultra-violet light. This is the type of light that can change the colour of your skin, and even cause skin cancer. There are different types of ultra-violet light, as you may have seen on the back of your suntan lotion bottle, known as UVA, UVB and UVC.

UVA is nearest to visible purple light, and that is the least dangerous. UVB is more likely to get you burnt and cause cancers. UVC is even more dangerous but at the moment not much of it reaches the surface of the earth because it is blocked by a gas called ozone high above us. Just as well!

Normal window glass blocks about a third of the UVA rays and around 97% of the UVB ones. This is about the same as wearing a factor 30 sunscreen, so you could still get burnt if you sat in the window for long enough. Car windscreens are made differently, though, with a thin layer of plastic between the glass, making it almost impossible for any damaging UV rays to get through to your skin.

Why is snow white when water is see-through?

We know that snow is just frozen water, *so how can two things made out of the same stuff look so different?*

It all comes down to the way that light travels and the tiny shapes inside the stuff we are looking at. Snow is formed high up in the clouds when water droplets freeze to form ice crystals of beautiful shapes. The ice crystals are also see-through, so light is travelling through them. But when the crystals stick together and fall in large enough numbers we get the carpet of white snow that we all hope to wake up to on a Christmas morning. Imagine how disappointing it would be if it wasn't white and we could still see the soil and roads through it!

Luckily, the surface of the snow at a tiny level is uneven and rough. So the light gets bent as it goes into the snow, and then bent again and again and again as it bounces around inside the snowflakes. Light is made up of all the colours of the rainbow, and all of these colours get bounced around in the same way. When they eventually find their way out of the snow and back to our eyes, all the colours are coming out together, and when our eyes see all the colours of the rainbow together, we see white.

The building blocks of water are different and the light doesn't bounce around in the same way as it does in snow. Because the light can travel straight through, we see water as see-through, or transparent.

Ice is usually see-through but it can look a bit white because as the water freezes you get small pockets of trapped air, or funny shapes forming between the water particles. These make the light bounce around a bit and make them look a little bit white.

Icicles are often very see-through because

they are formed by pure water freezing very slowly in neat layers, one on top of the other.

If you want the best chance of making very clear ice cubes, you should use boiled, distilled water as this means you'll have very pure water. Any other stuff in the water can change the colour you end up with.

Why is the sea salty?

If you've ever taken a mouthful of the stuff when hit by a wave, you know how the sea tastes. But where does all the salt come from? Rivers all over the world run into the seas, but they contain fresh water so that can't be the answer.

There are in fact three ways we get salt into the sea. Firstly, the rain flowing over the surface of the earth can dissolve salts and minerals in the soil and rocks and carry it to the sea. Secondly, the warmth of the sun heats the sea water and some of it turns to water vapour but the salt gets left behind, so the water left in the sea becomes even more salty. Finally, there are things called 'hydrothermal vents' which are like mini-underwater volcanoes. These erupt hot water that has gradually dripped through the rocks into the top part of the earth's crust, taking salts and minerals with it. When this water gets pumped back into the sea it has lots of salt from the crust of the earth in it.

The sea is on average about as salty as a glass of fresh water with a teaspoon of salt in it. The ocean contains loads of the stuff. In fact, if you could take out all the salt in the seas and spread it evenly over the surface of our planet, you would have a layer about 150 metres high – the same height as a 40-storey office block!

Some seas are saltier than others. The Red Sea between Africa and Asia is one of the saltiest oceans on earth because the hot temperature causes lots of the water to evaporate. You might also have heard of the Dead Sea, which is actually a large saltwater lake in Jordan. It is the lowest lake on earth, at 422 metres below sea level. Because it is about eight times as salty as the ocean, you can float on it really easily. All the salt in the water makes it very dense and the human body is a lot less dense, so we float. People often say this is the saltiest water on Earth but this isn't actually true. A lake in Antarctica – called Lake Don Juan – is the saltiest lake on Earth. Some scientists study the place to work out what life on Mars might be like because the conditions are similar.

How do nettles sting?

The leaves of nettles have amazing hollow hairs on them that are very, very easily broken because they are made of silica, the same stuff that makes glass! Inside these hairs is a mix of nasty chemicals that can annoy your skin. When you go to pick the nettle, or just brush past it, your hand (or leg) breaks off the end of these hollow hairs and the chemicals inside get injected into your skin. Your body reacts against the chemicals, making the sting hurt and itch.

Despite the well-known cure of rubbing a dock leaf on a nettle sting, there is no actual proof yet that this will help. Some people think that just rubbing the sting can stop the pain signals being passed on to the brain. Others say that the act of going looking for the dock leaf can stop you thinking about how much the sting is hurting you! Or perhaps it is just something called the 'placebo effect' where you think and believe the dock leaf will help you, so it does.

Nettles are in fact very good for you and are used as food in lots of places around the world. You can cook them in a similar way to spinach and they contain loads of vitamins, iron and protein. Once you cook the nettle, or chop it up, you will get rid of the stinging part and they are really healthy things to eat. The younger leaves have fewer stinging hairs so choose them if you ever want to try eating them raw.

Nettles are an incredibly successful plant that has done very well out of humans. You often find lots of nettles in places where humans used to live but have since left. They really love soil that is rich in human waste!

Can wasps smell fear?

The rare event of a picnic on a sunny day is often ruined by the appearance of nosy wasps. I have sat around with people completely happy to have wasps around them and it always seems to be those who are most scared (e.g. me!) who get bothered the most.

So can they really smell fear?
Humans do give off a scent that can be detected by others when they are afraid. Scientists have done tests using the sweat of people about to do their first sky dive and found that other humans can sense the difference between those who are scared and those who aren't. We give off smells as something called 'pheremones' and we know

that other people can smell these even when they aren't always aware of it. It is these pheromones that people say attract members of the opposite sex – although not usually the ones we make when afraid! So it is possible that wasps can pick up this smell, but no-one has shown that noticing the smell changes the way they behave. So are there other smells to blame?

Even when not scared, humans (particularly females) can give off something called acetic acid that has been shown to attract wasps. Acetic acid is what gives vinegar its taste, and is made when fruit or animal matter goes off. In humans it is made as an anti-bacterial treatment in the body. As this is what attracts wasps to

their usual food of over-ripe fruit, this could be one reason why they are so interested in humans. It is only when they get very close that they can work out you aren't in fact a decaying banana! By this time you have usually run around with waving arms and made them feel scared so they are more likely to defend themselves with a sting.

In wasp groups there are two types of job that a wasp might be doing. Some are 'foragers' who go out looking for food, and some are 'defenders' who are out and about looking for enemies of the wasp nest. It is usually the foragers looking for food who bother us most. It is quite rare to get bothered by wasps if there is no food around at all. So usually it is the food attracting the wasps, not your fear.

Killing or distressing a wasp is certainly the worst thing you can do as they give off pheromones of their own in a distress call to other wasps in the area. It is likely that killing or chasing one wasp may bring you more trouble. A better way to stop wasps bothering you is to hang a fake wasps' nest up where you are having your picnic. As wasps are very territorial in defending where they live, they will not normally come anywhere near another nest, and certainly not near enough to work out that it is fake. Has to be worth a try if it means eating your jam sarnies in peace and quiet!

Why do mosquitoes find some people more tasty than others?

You and your friend can be sitting out on the same balcony enjoying the summer evening, but the next day you could be covered in mozzie bites whilst your friend escapes bite-free. What is it about some people that attracts the little blood-suckers, and is there anything you can do to change your taste so you can put them off?

To some extent, how attractive you are to our biting friend depends on your blood type. There are four main blood groups, O, A, B and AB. Type O has been found to be the most attractive to mosquitoes. But it isn't just the blood group you have, it's how much you give off a noticeable signal of this blood group. And you can thank your parents for this signal as it is genetic, so you got it from them!

Another thing that can attract a bite is the amount of carbon dioxide you breathe out and the temperature of your body. This is bad news if you are pregnant as you are generally warmer and you breathe out a massive 21% more carbon dioxide than normal. At least this is a short-term condition that will pass, unlike having a certain blood group, which is pretty permanent.

More bad news for summer evenings is that drinking beer can also increase your chances of being bitten. This is partly due to the increase in your temperature but also because as your body digests alcohol it changes the way you release smells from your skin. There is a rumour that drinking tonic water can help because it contains a chemical called quinine that is used to treat malaria. This is unfortunately not true. The levels of quinine in tonic water are so low that you would need about five litres a day to have any effect. That's a lot of gin!

If you are someone who gets bitten a lot you may want to pay extra attention to the personal hygiene of your feet. Scientists do agree that some types of mosquito are attracted strongly to the aroma of smelly feet. In fact this has led to a rather bizarre idea of using a cheese called 'Limburger' as a mosquito trap. No prizes for guessing what the cheese smells like!

Chapter Four

FOOD AND DRINK

Why does popcorn pop?

And where does all the stuff come from that makes the finished popcorn about forty times bigger than it was before you started?

A corn kernel has lots of starch inside it, and a little bit of water, all contained within a hard shell. You have to heat the popcorn to very high temperatures – at least 200 degrees Celsius – to make it 'pop'. The water inside the corn turns to steam and this takes up much more space and puts the corn under very high pressure. The starch inside turns jelly-like and when the hard case finally gives way, the rapidly expanding steam blows the starch up into a thick, bubbly foam. As soon as the temperature cools, the foam goes hard, giving us the crunchy light texture we know and love.

Corn for popping has to have just the right amount of water inside. If you have too much water the corn pops too soon, before the starch inside has expanded, making the corn very hard. If there is not enough water, then there isn't enough steam to make the corn case pop open so you get those hard nuggets of un-popped corn that are so hard on your teeth.

But where does the popping sound come from?
When you have something changing shape very quickly you get a shock wave through the air. The rapid growth of the popcorn when it splits open squashes up the air around it very suddenly. A sound wave is made when you get a change in air pressure, so the sudden squashing up of the air sets up a Mexican wave movement through the air to our ears.

Be careful next time you tuck into a bag in the cinema. In February 2011 someone was shot dead in a cinema in Latvia after he was accused of eating his popcorn too loudly!

How does the 'widget' in a can of beer work?

If you take a widget out of the can it doesn't look very clever at all. It is simply a hollow, plastic ball, about 3 centimetres wide, with a tiny hole in it. So how can it make any difference to the beer?

As more and more people began to buy beer

in cans to drink at home, there became quite a demand for the 'draught' taste and this often related to the type of 'head' you could get on your beer. The 'head' of the beer depends on the amount of gas in the beer and how the bubbles of gas come out of it to create the foam on top. Stout beers like Guinness have a creamier head of foam that lasts longer than on most other beers.

Most beers are made fizzy using a gas called carbon dioxide. With very fizzy beers that don't require a big head of foam it is easy to pack them in cans and then add the carbon dioxide. When you open the can the bubbles of carbon dioxide come out of the beer and create bubbles which give you a short-lived small head. Guinness and other stouts have a different gas called nitrogen in them. Nitrogen bubbles are much smaller than carbon dioxide bubbles, which is why the Guinness head is a very creamy foam. It is harder to get these nitrogen bubbles out of the beer, which is where the widget comes in. The little ball is injected with liquid nitrogen and then sealed into the can. The nitrogen then turns from a liquid into a gas, creating a build-up of pressure inside the can. When you open the can, you release this pressure and a big stream of nitrogen gas comes out of the tiny hole in the widget. This starts a stream of bubbles through the stout and encourages more of the gas from the drink to come out and form even more bubbles. The tiny size of the bubbles means that the drink has a creamy texture and a longer lasting head, more like one that has been poured from draught in a pub.

The widget was originally a disc of plastic at the bottom of the can which was found to work well when served cold, but which tended to create too much foam in warmer weather. The plastic ball widget has been much better at all temperatures!

Why can't you make toast in the microwave?

Sometimes you just need toast quickly (see answer to the hangover question earlier!). But a toaster can seem to take forever, so maybe your microwave could do a quicker job?

There are different ways you can heat things up. Your toaster uses a type of heating called 'radiation'. If you peek inside when a toaster is switched on (be careful!) you will see some very thin wires glowing red hot. As well as the visible red light you can see, these wires are also giving out invisible heat called infra-red heat. This heat dries out the surface and makes it crispy and a bit sweeter where the bread has changed colour.

A microwave heats things in a very different way. The energy of microwave ovens makes the water inside your food vibrate very quickly. This movement creates a lot of rubbing together of the tiny parts inside the food, and this gives us the heat that cooks your food.

If you put a slice of bread in the microwave for a short time, the water in the bread gets very hot and turns to steam. The steam gets trapped in the bread and turns it very soggy. Not great with your marmalade! If you cook a slice of bread in the microwave for long enough it will eventually dry out and start to burn, but the bread will be dry and crunchy all the way through. Nothing like the delicious combination of crunchy outside and soft bread inside that we love from toast made in a toaster.

Why does toast always land butter side down?

Or even worse, strawberry jam side down, on your nice cream carpet…

It does seem that this happens more often than not, and in fact it can feel like the world is against you. How can I be so unlucky? Well, fear

not. The world is not picking on you, it is simply the laws of physics. No, wait, don't stop reading, it is honestly quite simple.

Most toast that falls to the ground was either on a plate or a work surface before it fell. This is important because it is rare for toast to be dropped flat onto the floor. Usually there is an element of sliding in the fall, and this is quite important (honestly – stay with me!). When the toast starts to slide off the plate, gravity begins to pull it towards the floor. As the back end of the toast is still on the plate, the toast tilts before it falls. This tilt means that the toast starts to spin. Once the toast has started spinning it will carry on spinning for as long as it is falling. Most toast only has about a metre to fall before the ground (or nice cream carpet) gets in the way. Because of the size of most bread slices, there is only time for it to complete half a spin before the ground stops it. As you usually carry your toast butter side up, this means it is more likely to end up butter side down, as that is where half a spin will take it.

You could make this less likely to happen by cutting your toast into four smaller squares. If these fall off your plate they are small enough to complete a whole spin before hitting the ground, so they are more likely to land butter side up!

Why do salty snacks make you thirsty?

Salt is an essential part of the human body and if you have ever bitten your lip or tongue or licked a cut, you'll know that blood has a slightly salty taste. In fact almost half a per cent of our body weight is salt. So what happens if you get too much?

As salt can be dissolved in water, the first

thing that happens in your mouth when you eat something salty is that it uses up a lot of saliva and this dries out your mouth. Although this is only a short-term effect it can make you feel you need more fluid to replace what you have used.

As the food moves through your body, the salt eventually gets into your bloodstream. Your blood is always slightly salty but if it becomes more salty, your body isn't happy and the water from within your cells comes out to try and wash out the salt. This means that your cells are left short of water, so they send an emergency signal to the brain saying you need to bring in more fluids.

The problem becomes worse if you eat salty snacks when having a drink or two. This means you get thirsty from the salt, have an (alcoholic) drink to quench the thirst (which takes away more water from the body), eat more snacks, get more thirsty, and so on, and so on. Just another good reason to have a glass of water to hand...

So should you ever drink sea water?
If you've ever seen films where people get stranded at sea, you'll know that even if you are dying of thirst you shouldn't think sea water can help because it is so salty. Drinking salt water pulls existing water from your body out

of your cells, and makes you wee more as your body tries to get rid of the salt. This makes you even more thirsty and worse off than you were before you had the drink!

Why do onions make you cry?

It never seems to happen to TV chefs, but I almost always end up in tears chopping onions for my kitchen creations. Why are they so nasty?!

Inside an onion are lots of cells containing chemicals. Whilst they are in your cupboard or fridge they are kept locked away and safe. If you cook your onions whole the chemicals turn into other things that aren't so painful to the eyes. But if, like most of us, you cut them when raw, watch out! Cutting the flesh of the onion breaks open the cells and lets out the chemicals. When these react with the air and the water in your eye they create a very nasty acid called sulphuric acid. This is very painful to your eyes, so the tear glands send a message to the brain to produce more tears to help wash it away.

Different types of onion can be better or worse for your tears. Generally, sweeter onions are not so bad, but very yellow onions, or ones grown in the autumn, are worse. It all starts with the onion taking in sulphur from the soil as it grows, so the type of soil can also have an effect.

You may have heard lots of ideas on how to chop onions without crying, could any of them help? Here are four ways that have some science behind them to suggest they could prevent your tears...

1. Cutting the onions underwater – not quite in a snorkel and scuba kit, but if you can manage to cut them under a running tap or a spray of water, the water should wash away the chemicals before they float up to your eye to irritate it.
2. Use a fan – if you can point a small fan at the onion it may blow the chemicals away from your eyes.
3. Put the onion in the freezer first – if you can chill it, the chemical reaction that causes the stinging will be much slower and if you're quick you may be able to get out of the danger zone by the time it kicks in.
4. Use swimming goggles – anything that can keep your eyes sealed from the air above the onion is going to help. Can't say you'll look that great though. Do it before your friends arrive for dinner perhaps...

Why does toothpaste make orange juice taste so bad?

A lot of people like to brush their teeth as soon as they get up before they have breakfast. If you've ever done this you might find that your orange juice has a really bitter taste to it. It isn't so much the mint taste of your toothpaste that does this but the foaming you get when you use most toothpastes.

Your tongue is able to detect five basic flavours: sweet, salty, bitter, sour and umami (a meaty taste – think Marmite!) but sometimes things can block the detectors and change the way something tastes.

Next time you pick up your toothpaste tube, have a look for something in there called 'Sodium Lauryl Sulphate' (or SLS). This is used in almost every toothpaste and its only job is to make a foam in your mouth. The foam doesn't clean your teeth any better but it makes your mouth *feel* cleaner! The problem is that this toothpaste foam blocks the sweet detectors on your tongue and actually helps the bitter ones work better. Orange juice is both sweet and bitter – so when your sweet detectors are blocked and your bitter ones are on high alert, the orange juice will taste really nasty. It can take some hours for the effect to wear off, so even a mid-morning juice can sometimes still taste 'dodgy'.

Why do some foods make you feel fuller for longer?

Hunger is partly to do with how your stomach feels and partly to do with what your brain is thinking. When your stomach is full your brain knows by feeling how stretched your stomach has become. This was tested out many years ago when a scientist swallowed a balloon attached to a pipe and inflated it to see if it made him feel full. Don't try this at home!

When your brain gets the message that your stomach is full, it produces a hormone called leptin. This is a really important hormone that affects whether you burn off your calories or store them all as fat. Foods full of fibre, such as oats, beans, broccoli and apples, help you fill your stomach and raise the level of leptin that your body produces. If you don't get enough sleep, the level of leptin can go down and trigger hunger, making you eat too much. Another good reason to get an early night!

Scientists have done lots of tests to see

which foods make you feel fullest for longest. The best foods for this were white potatoes (boiled, not fried!), eggs and oatmeal. Any foods that contain lots of water and air are also good, as you eat fewer calories per bite, but there's more stuff in your belly! Popcorn (lots of air) and apples (lots of water) are good at putting off hunger, and apples also have a special type of fibre in their skins that reduce your appetite.

You might have come across a 'GI diet' that is all the rage at the moment. The 'GI' stands for 'Glycaemic Index'. What?!

Well, basically it's a measurement of how quickly your blood sugar goes up after eating, and how high it goes. Some people suggest that the GI index of a food is directly linked to how full it makes you feel, but this isn't always true. The GI index is useful for people who are diabetic who have to control the amount of sugar in their system, but may not help other people battle the bulge.

Chapter Five

WEIRD AND WONDERFUL

Why do bright lights make you sneeze?

Well, perhaps not you personally but probably someone you know. About one in five of us has this condition where seeing bright light, particularly sunshine, can make you break out in a fit of sneezes. The effect is more noticeable when moving from a dark room into a very bright place, like when you come out of the cinema on a sunny day. But why should light make you want to sneeze at all?

A sneeze is usually the way your body gets rid of material in the nose that may be causing it to itch. The idea is that by clearing the nose of anything itchy, you also get rid of any nasties that might be hiding in your nose before they give you more trouble.

When your nose is irritated by something, a signal goes up to your brain along various paths and a message is then sent back to the nose, mouth, chest and eyes to prepare the body for the sneeze. When you see light, a similar signal goes from your eyes to your brain using nerves that are very close to the ones taking the 'sneeze' message.

The most likely reason why some people sneeze at bright light is that these two signals get mixed up somehow because the paths they take are very close. The route these signals take to the brain runs close to the skin just under your eyebrow. This is why plucking your eyebrows can make some people sneeze and why pressing down on your eyebrow can sometimes prevent you from sneezing (see 'Is it safe to sneeze when you are driving?').

How can chip fat power cars?

As petrol and diesel prices rise, and fossil fuels run low, everyone needs to find ways to do their bit to help. Some time ago there was a flurry of interest in the use of old cooking oil as an alternative to diesel. Can this really work? Are fish and chip shops all over the country sitting on a goldmine of fuel that we could all be using in our cars?

Let's think about what diesel is and work out why chip fat could be helpful. Diesel is made from crude oil (just like petrol) and the energy comes from the fact that it is made up of hydrogen and carbon, or hydrocarbon for short! Hydrocarbons come in lots of different types and chip fat is one example. Because chip fat is mostly vegetable oil (which comes from plants) the carbon dioxide given off when it burns is only the same as the amount the plant took in when it was growing. This means that overall the amount of carbon dioxide on the planet remains the same and we say that it is 'carbon neutral'. This makes it more planet friendly than petrol or diesel – and a little bit cheaper too!

Which way does water go down the plughole?

If you've got any friends 'down under', have you ever asked them which way the water in their sink and bath goes round when they empty it? Probably not. That would be weird. However you may have heard the theory that water spins one way down the plughole here in the northern parts of the world, and the opposite way below the equator.

A good question to ask might be why it spins at all. The shape of water going down the plughole is called a vortex and looks a bit like those twisters or tornados you see in America. A

vortex lets air travel up the centre of the tube whilst water flows down around the edges, and this shape is the fastest way to drain a sink. But a vortex needs a twist to get it set up, so where does this twist come from?

We all live on a massive spinning ball (the Earth) and because of this spin, when something falls it gets pushed in the same direction as the spin of the Earth. This force pushes one way in the northern parts of the world and the opposite way south of the equator. If you look at pictures of weather systems, you will see that huge hurricanes spin anti-clockwise in the north and clockwise in the south. This happens because as warm air moves from one place to another it gets a tiny push because of the spin of the Earth. As the hurricane is such a massive thing, all the little pushes across such a huge size of air creates a spiral.

This suggests that water down the plughole might do the same thing but this isn't true because your sink and bathtub are just too small for this effect to make any difference.

Why do you get pins and needles if you lie on your arm?

You know that weird – and sometimes slightly scary – sensation that you have lost all feeling in your arm? As you wake up the feeling changes from being completely numb to a prickling pins and needles feeling, sometimes joined by a burning sensation. It might not be very nice but it is in fact a brilliant way of protecting you from doing any serious damage to your body.

Taking your arm as an example, there are lots of nerves going up and down the arm that send different signals to the brain. They can tell you whether you are pushing against something, whether your skin is hot, and also help control your movement. If you put a lot of pressure on these nerves for a long time by sleeping on your arm, for example, you can stop these signals working. The pressure can also slow down blood flow to the area which makes the nerve cells behave in an unusual way.

Usually you get woken up at this stage by the strange feeling in your arm and you start to take action. As you rub your arm to try to bring back the feeling, different nerves come back to life at different times. Some stay asleep despite your efforts and others wake up so suddenly that they start firing random signals all over the place. This gives you the pins and needles feeling, that means things are getting back to normal.

Some of the nerves have thinner walls around them than others, and so they come back to life more quickly. The ones that send message about pain and temperature are thin, so usually the first sensation you will feel are the pins and needles, often followed by a slow burning feeling. The nerves that help you move your arm also recover quickly when you move, so you are usually able to move your arm about even though it may feel numb. If you ever touch your arm at this stage you may not be able to detect anything and that can make it feel like a 'dead' arm.

Although this is all a bit uncomfortable, it is done for good reason. If you put pressure on a limb for too long, and are not woken up, the damage to the nerves and cells in the arm could be permanent.

Can you think yourself thin?

It sounds too good to be true, but on a cold wet night when you really can't face going out for a run, could you just sit and think really hard to burn some of those chocolate calories?

Even without doing anything, your brain needs calories to keep you alive. You would be using about a tenth of a calorie every minute, just by living. But if you work your brain a bit harder by reading a book, the level can go up to about one and a half calories every minute, so about 90 calories an hour. A small (125 ml) glass of red wine is about 85 calories, so if you make your wine last you could treat yourself – as long as you think really hard for the full 60 minutes!

This sounds pretty good but when compared to walking (4 calories a minute) or jogging (10 calories a minute) it's not a lot. It isn't going to be enough on its own to keep you fit.

Why should thinking use energy anyway?

Well, the brain is made up of lots of cells called 'neurons' that send messages to each other. They make chemicals called 'neurotransmitters' to send these messages and this needs glucose (a type of sugar) and oxygen. The neurons get these supplies from the blood, and the blood in turn is getting sugars from the food you eat. This is why a session of deep thought can use up a few of those calories you have eaten.

Do ghosts exist?

In 2009 a survey was done that found that 4 out of 10 people in the UK believe in ghosts. But in scientific terms is it enough to say something is true just because a large number of people believe it?

No.

After all, everyone used to believe the world was flat until there was enough proof that it wasn't. A belief in ghosts is very much a personal thing, but if they really don't exist, how can we can explain some of the spooky things that go on? The ghostly figures, the hairs on the back of your neck standing up, the unsettled feeling in your stomach? Can science shed any light on the subject?

Some years ago an engineer called Vic Tandy was working on his own in his lab (they always are, aren't they?!) and he had a ghostly experience. It was a very old building and out of the corner of his eye he saw a flash of white that appeared to move across the room, and he started to feel very odd. The next evening he was alone again and fixing a fencing foil, which is a long and very thin blade. He had it held tightly in a clamp on the bench with the blade pointing upwards and suddenly the blade began to swing about.

Despite being pretty scared he decided to try and measure what could be making the blade move. Knowing that sound waves could cause movement in air and objects, he decided to get out a machine that could measure sound. He knew he couldn't *hear* any sound but the machine showed a high level of sound below the level humans can hear. Very, very low-pitched sounds that we can't hear are called 'infrasound'. In an empty building when everything else is still, moving machines, such as fans or air conditioners can create these infrasound waves. Even something as simple as an open window with the wind blowing outside could cause the effect. But why would this 'silent' sound wave make you see and feel things that aren't there?

Well, every part of your body – depending on size and shape – will tend to vibrate if the right sound is played. Your eyeballs like to wobble at about 19 times a second so if a sound wave matches that, it would create very small movements in your eyes. This movement means your eyes could fire off false messages to the brain that then thinks you are seeing movement that isn't there. Possibly an explanation for the ghostly images people report seeing?

Not only this, but your stomach has a low vibration rate too, so infrasound can make you feel a bit sick or uneasy. Scientists have tried this out by adding infrasound to music and, although you can't hear it, people have reported feeling very uneasy when the infrasound is playing. Quite a lot of 'haunted' houses have been show to have infrasound in them, and it seems that if the effect is strong enough it can even make physical objects move (like with Vic's fencing blade). Perhaps this is even an explanation of poltergeist activity?

I think many people like to believe in ghosts because the alternative idea of a life ending when we die is too hard to accept. Science alone certainly can't give us all the answers on this one yet...

What is your sixth sense?

To some people a sixth sense might suggest some spooky abilities to predict the future or read minds, but there is a real sense we have that is often called our sixth sense and it's absolutely essential to our everyday lives. Without it we couldn't drive a car, type on a computer, or send text messages on our phones, and even walking would take a massive amount of concentration.

It is the sense of knowing where all the bits of our bodies are without having to look at them! You can test out this sense for yourself by putting both hands above your head, closing your eyes, then trying to touch each of the fingers of one of your hands with the index finger on the other hand. You should touch your nose in between each finger to make it a more difficult task. If you find it very hard you can wriggle your fingers which makes it easier to find them.

When you do this, you can't see where your fingers are, and they aren't touching anything, so you are not using any of the five basic senses. Your brain is able to steer your hand because nerves in the tendons and muscles can tell you how stretched out your arm and hand is. This is a vital sense to have when driving a car because it means you do not need to look at your hands on the steering wheel or your feet on the pedals to know exactly where they are. Sending a text with your thumb whilst you look at the phone screen is a similar example.

Can stress make you bald?

In short, yes it can.

Your hair has three stages of growing: active growth, transition and finally a resting phase. About 85% of your hair is in the growing phase at any one time, with the rest in transition or resting. An increase in stress – from a physical injury or emotional problem – can make more hairs suddenly enter the resting stage. It's a way for your body to focus on the more important problems you may be having.

This stress-related type of hair loss happens about three months *after* the stressful event, so you might not make a connection between the two things. The good news is the hair will regrow once the stress levels have dropped. This very often happens to pregnant women about three months after having the baby. The drop in hormones after having the baby sends the hairs into resting period, and then three months later a larger than normal number of hairs all fall out together.

Different hair cells across your body have different maximum times of growth which also explains why the hair on your arms, and… er… other bits, will never grow long enough to put in an impressive French plait. Hair grows about 15 centimetres a year and most people have a growth time of 30–45 days for the hair on their arms and legs, meaning that however thick and bushy it may seem, it probably won't grow much beyond a centimetre in length. This difference in growth time is also why some people can grow their hair very long and others struggle to get it past shoulder length.

About the Author

Wendy Sadler is the founding Director of *science made simple* – the award-winning company which specialises in making science inspirational and fun. She has written 19 books for children, presented science on TV shows including *Tomorrow's World* and the *Alan Titchmarsh* show, and is a contributor on BBC Radio Wales. Her company's science shows have toured in 22 countries and appeared at the Edinburgh Festival Fringe. Wendy is currently employed by Cardiff University.